Hla Myo Tun
Thant Zin Win
Yusui Nakamura

Film Formation and Characterization of Undoped ZnO by Mist-CVD System

Hla Myo Tun
Thant Zin Win
Yusui Nakamura

Film Formation and Characterization of Undoped ZnO by Mist-CVD System

LAP LAMBERT Academic Publishing

Imprint

Any brand names and product names mentioned in this book are subject to trademark, brand or patent protection and are trademarks or registered trademarks of their respective holders. The use of brand names, product names, common names, trade names, product descriptions etc. even without a particular marking in this work is in no way to be construed to mean that such names may be regarded as unrestricted in respect of trademark and brand protection legislation and could thus be used by anyone.

Cover image: www.ingimage.com

Publisher:
LAP LAMBERT Academic Publishing
is a trademark of
Dodo Books Indian Ocean Ltd. and OmniScriptum S.R.L publishing group

120 High Road, East Finchley, London, N2 9ED, United Kingdom
Str. Armeneasca 28/1, office 1, Chisinau MD-2012, Republic of Moldova, Europe
Managing Directors: Ieva Konstantinova, Victoria Ursu
info@omniscriptum.com

Printed at: see last page
ISBN: 978-3-659-63186-3

Zugl. / Approved by: Kumamoto, Kumamoto University, Post-doctoral Research Report, 2015

Dedication

To Myat Su Nwe with Love

Acknowledgements

This research work is made possible through the help and support from everyone, including: parents, teachers, family, friends, and in essence, all sentient beings. Especially, please allow me to dedicate my acknowledgment of gratitude toward the following significant advisors and contributors:

First and foremost, I would like to thank **Professor Dr. Yusui NAKAMURA** (Professor, Graduate School of Science and Technology, Kumamoto University) for his most support and encouragement. He kindly read my research work and offered invaluable detailed advices on grammar, organization, and the theme of the research work.

The authors sincerely acknowledge **Dr. Hiroshi SHIRAKAWA** for valuable discussions. This work is supported for EEHE Project from Japan International Cooperation Agency (JICA).

Finally, I sincerely thank to my parents, family, and friends, who provide the advice and financial support. The product of this research work would not be possible without all of them.

Abstract

ZnO thin films were deposited on sapphire substrate by mist chemical vapor deposition (mist-CVD) with different flow rate of carrier gas. This is a simple and low cost method for large-area deposition system. At first, 60mL of Zinc Chloride solution and Nitrogen were used as sources, and the crystal growth was performed at various temperatures with 8 L/min flow rate of Nitrogen gas. In the second experiment, zinc chloride solution was used as sources, and the crystal growth was achieved at the growth temperature of 600°C and various flow rates of Nitrogen gas. The X-ray diffraction (XRD) spectrum was performed, and the photoluminescence spectra proved near-band-edge emission and strong deep-level emissions. In this work, we obtained the optimum condition for crystal growth of ZnO on m-plane sapphire, where XRD θ-2θ single peak at m-plane ZnO. Based on the results, it is confirmed that ZnO thin films can be a potential candidate for blue light emitting diodes (LEDs).

Table of Contents

Chapter 1

Introduction

1.1. Overview

LED technology initiates in the same art of engineering that gave us mobile phones, computers andall modern electronics equipment based on quantum phenomena. Extending semiconductor devices to new compoundshas produced great benefits to human life, as exemplified bymodern optoelectronic and high-speed electronic devices forcommunications attained by Zinc Oxide, Blue LED etc. Engineers use a blue LED to excite some kind of fluorescent chemical in the bulb. That converts the blue light to white light.To harvest these advantages in real devices, a reliable technique for fabricating p-type doping needs to be established.Zinc oxide has substantial advantages including largeexciton binding energy, as demonstrated by efficient excitonic lasingon optical excitation.

Figure.1.1. Blue LEDs

The optical properties of undoped Zinc Oxidewere observed by experimenting with various substrate temperatures with Mist Chemical Vapor Deposition (Mist-CVD) technique was observed in this work. The observation from optimal values for crystal growth and XRD θ-2θ single peak at m-plane Zinc Oxide could be supported to implement the blue emission layer of Blue LEDs.

Characterizations for undoped Zinc Oxide were performed by Scanning Electron Microscope (SEM), He-Cd Laser for Photoluminescence Measurement, X-ray Diffraction (XRD) θ-2θ and film thickness measurement in Oxide Semiconductor Device Laboratory.

1.2. Research Motivation

Recently, oxide semiconductor materials have been of great interest due to application for optical devices such as light emitting diodes (LEDs) and laser diodes (LDs). Zinc Oxide (ZnO) has enthralled extensive consideration due to its superior physical properties and wide technological applications [1-3]. The wide direct-band gap of ZnO is 3.37eV and it has a large exciton binding energy of 60meV, which tasks efficient excitonic emission processes at room temperature and enables devices to rationale at a low threshold voltage. ZnO (as a group-II oxide) proves vast guarantee for applications in blue/UV light emitters and photodetectors, over and above transparent electronics, chemical sensors, spintronics, and varistors. Various techniques, such as magnetron sputtering, reactive evaporation, pulse laser deposition (PLD), metalorganic chemical vapor deposition (MOCVD) and molecular bean epitaxy (MBE) can be constructive for ZnO thin films deposition.

1.3. Objectives of Research

For electronic applications, the attractiveness of ZnO lies in having high breakdown strength and high saturation velocity. ZnO also affords superior radiation hardness compared with other common semiconductor materials, such as Si, GaAs, CdS, and even GaN, enhancing the usefulness of ZnO for space applications. The objectives of research are

- To investigate the crystal growth of ZnO on m-plane sapphire
- To scrutinize the optical properties based on the various characterization methods

1.4. Research Direction

In this research work, quantitative comparison for film formation and characterization of undoped ZnO on m-plane sapphire by mist chemical vapor deposition (mist-CVD) will be focused. The two experiments are performed at various temperatures with 8 L/min flow rate of Nitrogen gas and at the growth temperature of 600°C and various flow rates of Nitrogen gas.

1.5. Contribution of Research

We have developed a mist-CVD system as a talented technique that allows better controllability in film deposition at low cost with a simple system and low energy consumption [4-5]. In this method, zinc chloride (zinc compound) solution is ultrasonically atomized to form mist particles of the solution, and the particles are consequently transferred by a carrier gas of nitrogen onto the heated sapphire substrate, forming ZnO film by pyrolysis and chemical reactions. The fundamental concept may resemble that of spray pyrolysis, but the main difference is in particle volume and the merits of treating mist particles like nitrogen gas. We intend to measure the front side (Inflow Side), center and back side (Exhaust Side) of deposited thin films of ZnO.

1.6. Scope of Research

In this research works, we report that ZnO layers deposited on m-plane sapphire substrate by mist chemical vapor deposition (mist-CVD) with different flow rate of carrier gas at 600°Cexperiments. The optical and structural properties of un-doped ZnO layer are characterized by scanning electron microscopy (SEM), photoluminescence (PL), film thickness measurement and X-ray diffraction (XRD). Based on the results, it is confirmed that ZnO films have single orientation of crystal.

6

1.7. Originality of the Report

The originalities of the report are:

- Finding the evidence of crystal growth of ZnO on m-plane sapphire
- Measuring the characterization of developed thin film with SEM, PL, XRD and thickness measurement
- Proving the optical properties of the developed ZnO thin film

1.8. Organization of Research Report

This research report is organized into six chapters. Chapter 2 discussed the crystal growth of ZnO and reviews current state of the art implementations. Chapter 2 also gives some background information regarding the film formation of ZnO. Various investigations on ZnO thin film that have been developed throughout the years are also compared, and film formation approaches are introduced.

Chapter 3 focused on the film formation and characterization of undoped ZnO on m-plane sapphire by mist chemical vapor deposition (Mist-CVD).

Chapter 4 presents the experimental setup for mist chemical vapor deposition (Mist-CVD) system.

Chapter 5 mentions the experimental results on the film formation and characterization of undoped ZnO on m-plane sapphire by mist chemical vapor deposition (Mist-CVD). The application areas of the proposed research work were also described in this chapter.

Chapter 6 discusses the research outcomes of the proposed research work.

Chapter 2

Literature Survey

2.1. ZnO Polytype Structures

The tetrahedrally coordinated bonding geometry determines the ZnO crystal structure. Each zinc ion has four oxygen neighbour ions in a tetrahedral configuration and vice versa. This geometrical arrangement, which is well known from, for example, the group-IV elements C (diamond), Si, and Ge, is also common for II–VI and III–V compounds. It is referred to as covalent bonding, although the bonds may have a considerable degree of polarity when partners with different electronegativity are involved. The tetrahedral geometry has a rather low space filling and is essentially stabilized by the angular rigidity of the binding sp^3 hybrid orbitals. In a crystal matrix, the neighbouring tetrahedrons form bi-layers in the ZnO case, each one consisting of a zinc and an oxygen layer. Generally, this arrangement of tetrahedrons may result either in a cubic zinc-blende-type structure or in a hexagonal wurtzite-type structure, depending on the stacking sequence of the bi-layers.

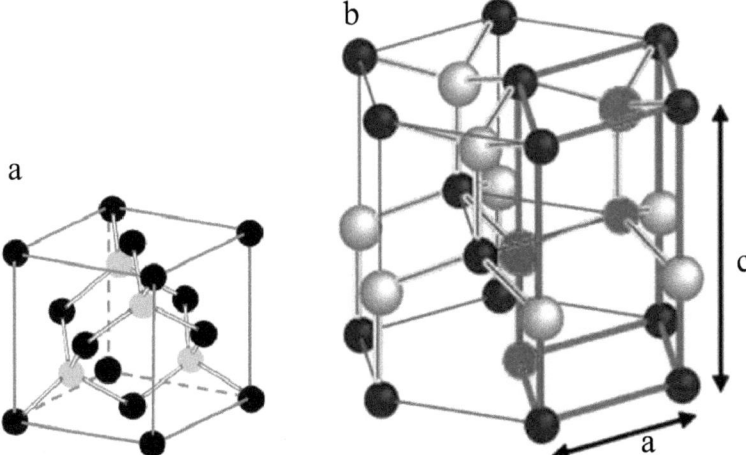

Figure. 2.1 The cubic zinc-blende-type lattice (a), and hexagonal wurtzite-type lattice (b). In the wurtzite lattice, the atoms of the molecular base unit (2×ZnO) are marked by *red* full circles and the primitive unit cell by *green* lines

The zinc-blende structure is shown in Figure 2.1a. It may be regarded as an arrangement of two interpenetrating face-centered cubic sub-lattices, displaced by 1/4 of the body diagonal axis. The bonding orbitals are directed along the four body diagonal axes. Note that the cubic unit cell is not the smallest periodic unit of a zinc-blende crystal, i.e. it is not a primitive unit cell [11]. The primitive unit cell of zinc-blende is an oblique parallelepiped and contains only one pair of ions, in our case, Zn^{2+} and O^{2-}. In group theory, this lattice is classified by its point group T_d (Schoenflies notation) or $\bar{4}3m$ (international notation) and by its space group, denoted as T_d^2 or $F\bar{4}3m$, respectively [12]. In contrast to the cubic geometry, the hexagonal wurtzite lattice shown in Figure. 2.1b is uniaxial.

2.2. Investigation of Previous Research Works

Toru Aokia and Yoshinori Hatanaka [13] have presented ZnO diode fabricated by excimer-laser doping. A ZnO diode was fabricated by using a laser-doping technique to form a p-type ZnO layer on an n-type ZnO substrate. A zinc-phosphide compound, used as a phosphorous source, was deposited on the ZnO wafer and subjected to excimer-laser pulses. The current–voltage characteristics showed a diode characteristic between the phosphorous-doped p-layer and the n-type substrate. Moreover, light emission, with a band-edge component, was observed by forward current injection at 110 K in that work.

Y.R. Ryu, W.J. Kim, H.W. White [14] discussed on the fabrication of homostructural ZnO p-n junctions. For the first time homostructural zinc-oxide (ZnO)-based p-n junctions were successfully fabricated. As-doped ZnO films were used for the p-type sides and Al-doped ZnO films for the n-type sides of p-n junctions. ZnO films have been deposited on p-type GaAs substrates by pulsed laser ablation in an ambient gas of oxygen to reduce oxygen vacancies. The ambient gas pressure was 40 m Torr and the substrate temperature was 350-450°C. The current-to-voltage (I-V) curves of the ZnO films very clearly show the characteristics of p-n junctions. This is the first demonstration of a p-n junction fabricated entirely with ZnO.

9

P. Zu Z.K. Tang, G.K.L. Wong, M. Kawasaki, A. Ohtomo, H. Koinuma and Y. Segawa [15] mentioned ultraviolet spontaneous and stimulated emissions prom ZnO microcrystallite thin films at room temperature. Room-temperature free exciton absorption and luminescence were observed in ZnO thin films grown on sapphire substrates by the laser molecular beam epitaxy technique. At moderate optical pumping intensities, an exciton-exciton collision induced stimulated emission peak was observed at 390 nm. The existence of this peak was related to the presence of closely packed hexagonally shaped microcrystallites in these films. Stimulated emission due to electron-hole plasma recombination process was also observed at higher pumping intensities.

Y.R. Ryu, S. Zhu, D.C. Look, J.M. Wrobel, H.M. Jeong, H.W. White [16] expressed the synthesis of p-type ZnO films. The p-type ZnO obtained by arsenic (As) doping is reported for the first time. Arsenic-doped ZnO (ZnO:As) films have been deposited on (0 0 1)-GaAs substrates by pulsed laser ablation. The process of synthesizing p-type ZnO:As films was performed in an ambient gas of ultra-pure (99.999%) oxygen. The ambient gas pressure was 35 m Torr with the substrate temperature in the range 300-450°C. ZnO films grown at 400°C and 450°C were p-type and As is a good acceptor. The acceptor peak was located at 3.32 eV and its binding energy is about 100 meV. Acceptor concentrations of As atoms in ZnO films were in the range from high 10^{17} to high 10^{21} atoms/cm^3 as determined by secondary ion mass spectroscopy (SIMS) and Hall effect measurements.

Bixia Lin and Zhuxi Fu [17] explored the green luminescent center in undoped zinc oxide films deposited on silicon substrates. The photoluminescence (PL) spectra of the undoped ZnO films deposited on Si substrates by DC reactive sputtering have been studied. There are two emission peaks, centered at 3.18 eV (UV) And 2.38 eV (green). The variation of these peak intensities and that of the I –V properties of the ZnO/Si heterojunctions were investigated at different annealing temperatures and atmospheres. The defect levels in ZnO films were also calculated using the method of full-potential linear muffin-tin orbital. It is concluded that the green emission corresponds to the local level composed by oxide antisite defect O_{Zn} rather than oxygen vacancy V_O, zinc vacancy V_{Zn}, interstitial zinc Zn_i , and interstitial oxygen O_i.

Chapter 3

Mist Chemical Vapor Deposition (mist-CVD)

3.1. Procedure for Mist-CVD System

The apparatuses for mist-CVD system are illustrated in Figure.3.1. We apply these apparatuses and follow the step-by-step procedures of the followings. Firstly, we have to clean all equipment (Etching with pure water) such as chamber and so on. We have to make quartz vibration system using pure water under the bottle with Zinc Chloride ($ZnCl_2$) 60ml. We have to setup the chamber and bottle for cooling water system (specify the Nitrogen gas flow rate). (*Various flow rate of nitrogen gas such as 6l/min, 8 l/min and 10 l/min*) And then we have to turn ON the furnace and heat wait for 20 minutes of preparation time and 30 minutes for deposition time. (*Various temperature range from 550°C, 600°C and 650°C*) After heating, later 50 minutes, we have to wait for down stage of temperature 300°C. And we have to make etching the chamber and measure the remaining quantities of Zinc Chloride ($ZnCl_2$). We have to wash the bottle and vibration system.

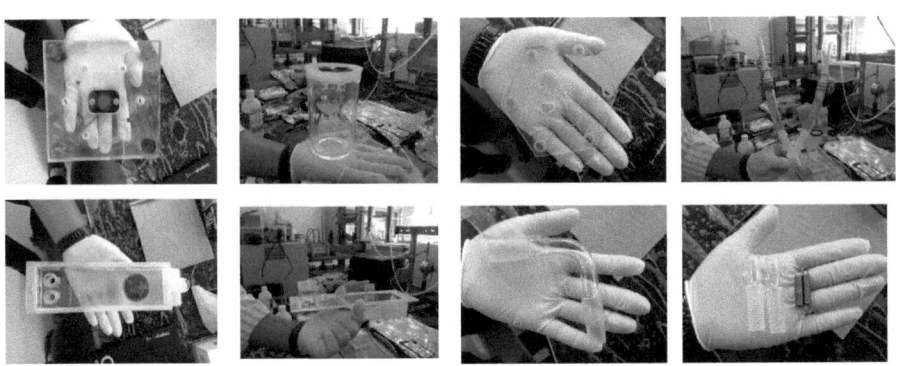

Figure.3.1. Apparatuses for mist-CVD System

3.2. Preparation for Furnace

The furnace for mist-CVD system is shown in Figure.3.2. There are four main portions in this furnace such as outer frame, outer cover, sliders and temperature control for mist-CVD system. Firstly, we have to make the setting of necessary temperature for mist-CVD system. We have to do experimental setup for mist-CVD system and the process is discussed in the next chapter.

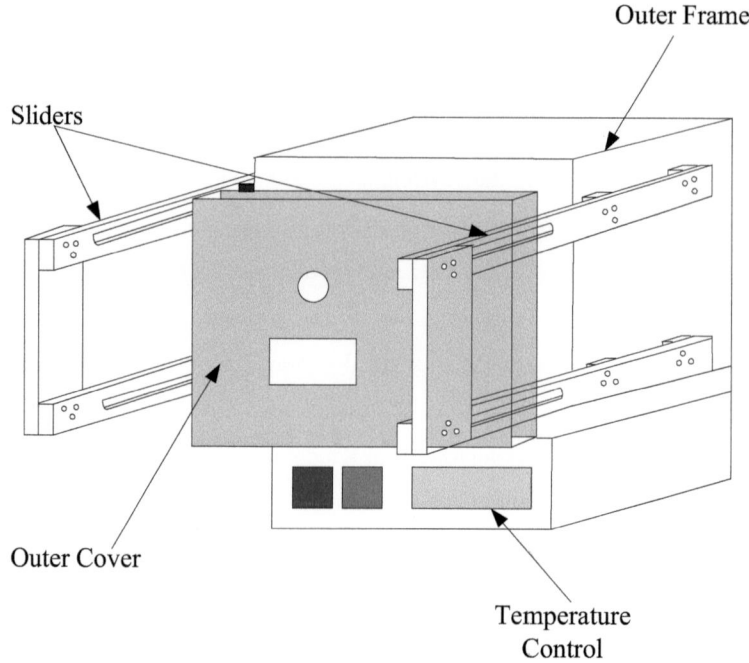

Figure.3.2. Furnace for Mist –CVD System

Chapter 4

Experiments

4.1. Experimental Setup

Figure 4.1 shows the experimental setup of mist-CVD system for crystal growth of undoped ZnO films were grown on m-plane sapphire with different flow rate of carrier gas. Deposition of ZnO thin films was carried at the substrate temperatures of 600°C with various gas flow rates of 6 L/min, 8 L/min and 10 L/min and 60 mL of zinc chloride solution.

The photoluminescence spectra were verified by using He-Cd Laser which has an excitation wavelength of 325nm. After the thin film was etched by mixed solution of CH_3COOH solution, phosphoric acid and pure water and the film thickness was measured by KLA-Tencor. The surface of film morphology was investigated with JEOL JSM7600F (SEM). The X-ray Diffraction (XRD) scan in the $\theta/2\theta$ mode was performed to determine the film orientation perpendicular to the film surface.

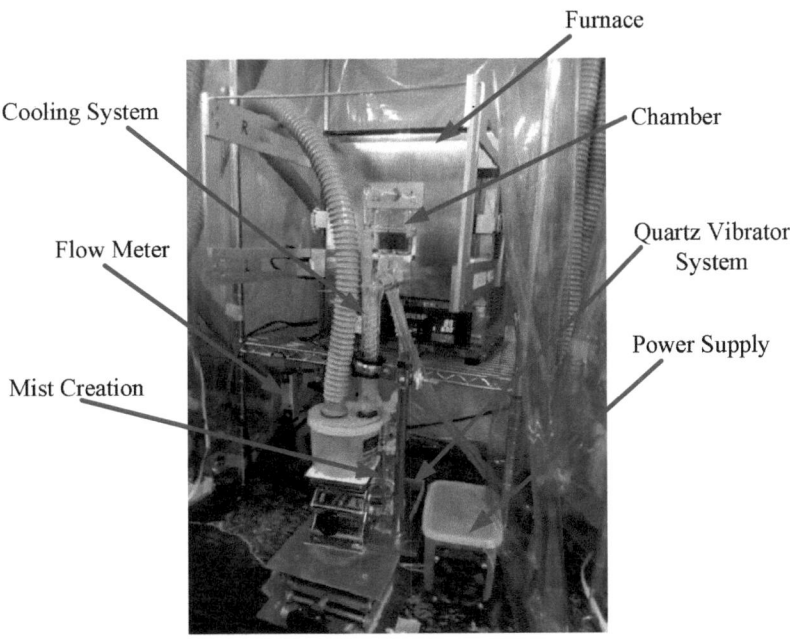

Figure.4.1. Experimental Setup

4.2. Working Stage

After setting the necessary temperature for two experiments, we took several times exactly one hour for mist-CVD process. Firstly, we took preparation time for twenty minutes with high temperature. And then, we took thirty minutes for mist process. After that, we have ten minutes for cooling time. Figure.4.2 shows the operating condition of Mist-CVD System.

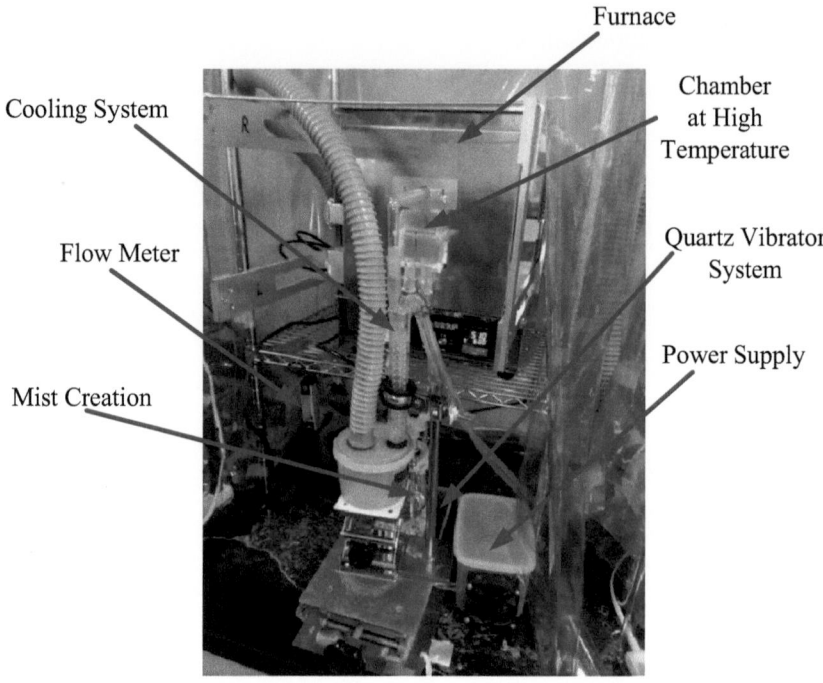

Figure.4.2. Operating Condition of Mist-CVD System

Chapter 5

Experimental Results

5.1. Results and Discussions on First Experiment

The deposition condition for first experiment is given in Table.5.1.

Table.5.1. Deposition Conditions for First Experiment

Solution	Zinc Chloride aqueous solution (0.1 mol/L)
Deposition Temperature	500, 600 and 650°C
Flow Rate	8 L/min
Substrate	m-plane Sapphire
Solution Amount	60ml

Figure.5.1 exhibits the photographs of deposited thin films for ZnO with various temperatures such as 600°C, 650°C and 550°C with 8 l/min flow rate of nitrogen gas.

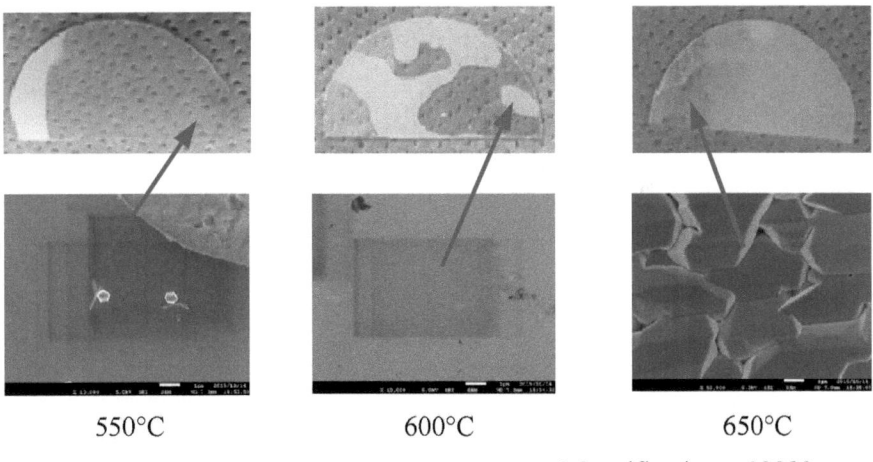

| 550°C | 600°C | 650°C |

Magnification :x10000
Position:Front (about 15 mm)

Figure.5.1. Crystal Growth of ZnO on m-plane Sapphire (a) 8 L/min with 600°C (b) 8 L/min with 650°C (c) 8 L/min with 550°C

The photoluminescence spectra were verified by using He-Cd Laser which has an excitation wavelength of 325nm. We have to check the adjustment of apparatus for PL measurement and open the computer which is connected to the Laser Setting the sample on the sample base. And then we have to adjust the integral time for PL measurements and get the spectrum of intensity (a.u) with respect to wavelength (nm). After that we have to save the spectrum of each sample by .raw format by using OOIBase32 software and convert the .txt format from . Raw by using plot32 software.

In order to explore the optical properties of ZnO homojunction deposited on sapphire, the PL measurements were performed at room temperature and the results are shown in Figure.5.2. A strong near-band-edge (NBE) emission peak was observed at 3.263eV, while the deep level emission related to structural defects was very weak. The results point out that ZnO film on sapphire substrate is optically high quality.

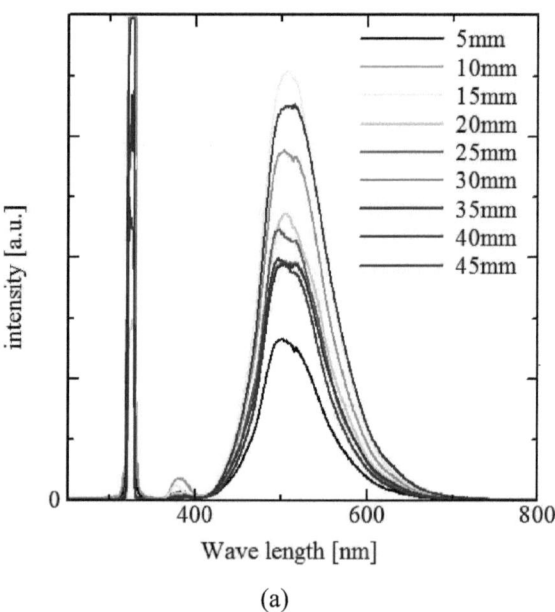

(a)

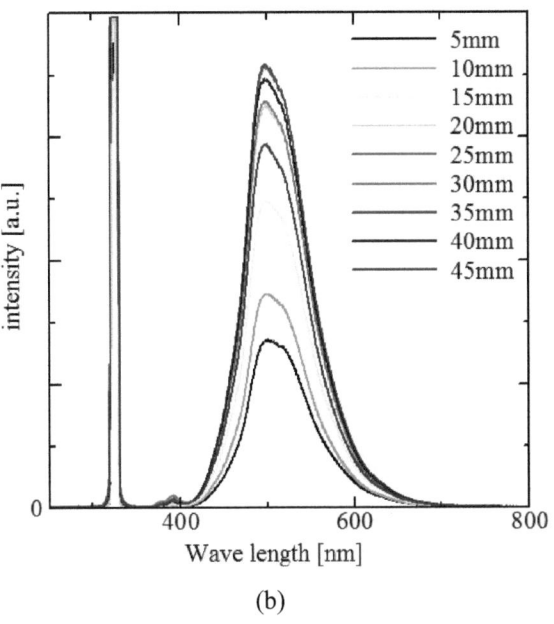

(b)

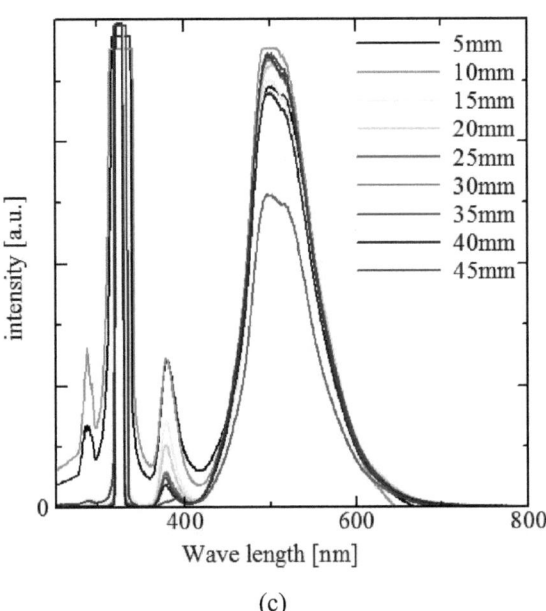

(c)

Figure.5.2. Photoluminescence Spectrum for (a) 8 L/min with 600°C (b) 8 L/min with 650°C (c) 8 L/min with 550°C

We have to check the amount of Nitrogen Gas (N2) for measurement and switch on computer for operation. And then we have to do setting the sample on the sample base and place sample in JEOL JSM 7600F Field Emission Electron Microscope. After that we have to do vacuum in JEOL JSM 7600F Field Emission Electron Microscope and measure the morphological features of the surface of sample by 10K, 20K of x level for front, center and back position of sample. We have to check the grain boundary level whether the sample is good or bad and compare and analyze the morphological features of the surface of sample.

The defect emission is strong in all samples. The Band edge emission is located at 380nm and the defect emission is found between 500 and 600nm.

The surface morphologies observed by SEM for (a) 8 L/min with 600°C (b) 8 L/min with 650°C (c) 8 L/min with 550°C are illustrated in Figure.5.3, 5.4 and 5.5, respectively. At 550°C and 600°C, the front side of the sample has almost no grain boundary. All samples have good crystalline condition.

(a) Back 10K

(b) Back 20K

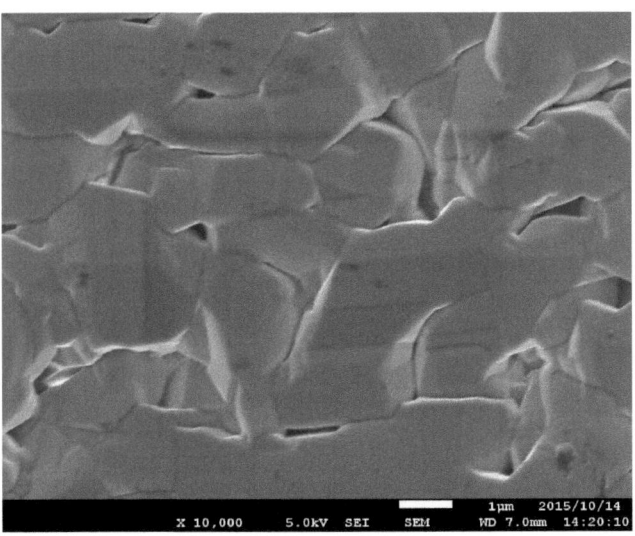

(c) Center 10K

Figure 5.3 (a) and (b) illustrates the surface morphology for the position of 10000 μm and 20000μm of back side of the sample for 8 L/min with 600°C. There are some grain boundaries in this side.

20

Figure 5.3 (c) and (d) demonstrates the surface morphology for the position of 10000 μm and 20000μm of center of the sample for 8 L/min with 600°C. This is the same condition of back side of the sample.

(d) Center 20K

(e) Front 10K

Figure 5.3 (e) and (f) shows the surface morphology for the position of 10000 µm and 20000µm of front side of the sample for 8 L/min with 600°C. There are no grain boundaries in this portion. This region is useful for emission layer of LEDs.

(f) Front 20K

Figure.5.3. Surface Morphologies observed by SEM (8 L/min with 600°C)

(a) Back 10K

Figure 5.4 (a) and (b) explains the surface morphology for the position of 10000 μm and 20000μm of back side of the sample for 8 L/min with 650°C. There are high grain boundaries in this side.

(b) Back 20K

(c) Center 10K

Figure 5.4 (c) and (d) proves the surface morphology for the position of 10000 μm and 20000μm of center of the sample for 8 L/min with 650°C. This is the same condition of back side of the sample.

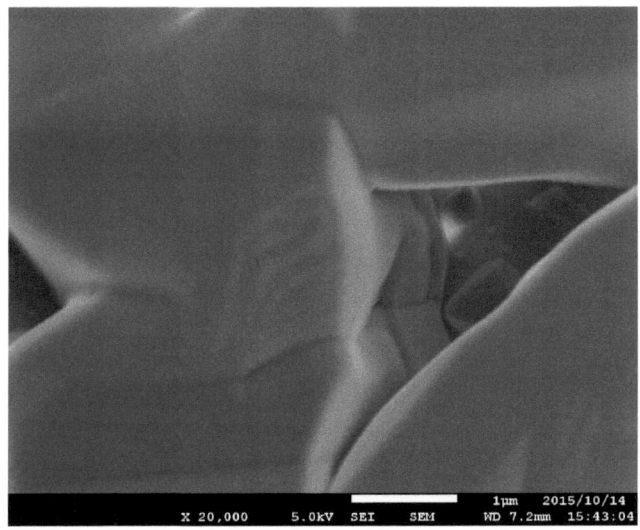

(d) Center 20K

(e) Front 10K

Figure 5.4 (e) and (f) confirms the surface morphology for the position of 10000 μm and 20000μm of front side of the sample for 8 L/min with 650°C. This is the same condition of all side of the sample.

(f) Front 20K

Figure.5.4. Surface Morphologies observed by SEM (8 L/min with 650°C)

(a) Back 10K

Figure 5.5 (a) and (b) shows the surface morphology for the position of 10000 μm and 20000μm of back side of the sample for 8 L/min with 550°C. There are high grain boundaries in this side.

(b) Back 20K

(c) Center 10K

Figure 5.5 (c) and (d) illustrates the surface morphology for the position of 10000 µm and 20000µm of center of the sample for 8 L/min with 550°C. There are small grain boundaries in this side.

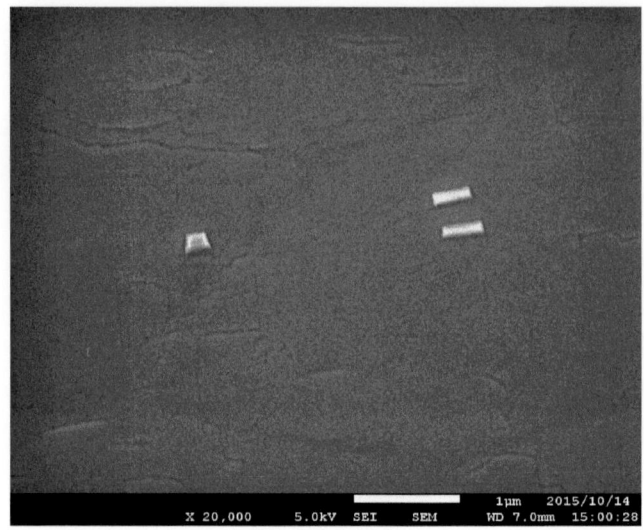

(d) Center 20K

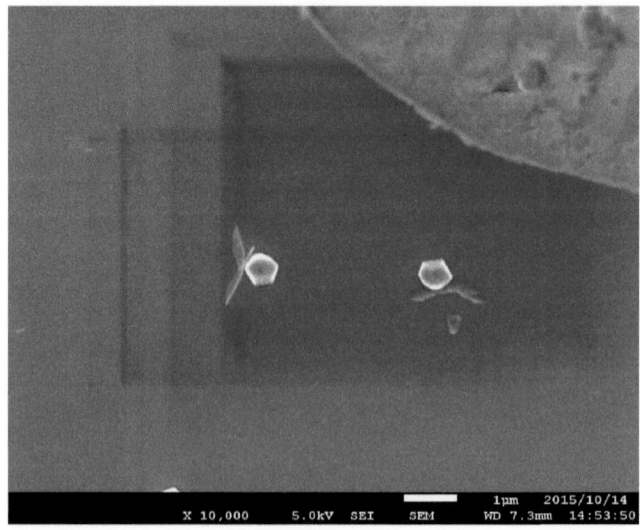

(e) Front 10K

Figure 5.5 (e) and (f) demonstrates the surface morphology for the position of 10000 µm and 20000µm of front side of the sample for 8 L/min with 550°C. There are small grain boundaries in this side.

(f) Front 20K

Figure.5.5. Surface Morphologies observed by SEM (8 L/min with 550°C)

After checking the surface morphologies, we have to characterize the crystal growth of all samples. We have to check the certain gas level of [Ar+Ch₄] about 0.07MPa and 10Mpa and switch on the computer and XRD machine. And we have to open the XG-RINT2500(Cu) or XG Software and specify the setting of voltage, current and power level for measurement [20kV,10mA and 0.2kW]. After that we have to prepare the sample for XRD measurement (front, center and back) with standard case and to specify the setting level for [start,stop, step and speed as 20, 80, 0.05 and 10) and [counts and 2θ/θ]. We have to use PXDL2 software for interpretation of XRD Spectrum and analyze the (100 for m-plane,001 for c-plane,110 for a-plane and 102 for r-plane) from the XRD spectrum. Finally we have to change the file type from .raw to .txt.

The XRD spectrum for (a) 8 L/min with 600°C (b) 8 L/min with 650°C (c) 8 L/min with 550°C are shown in Figure.5.6, Figure.5.7 and Figure.5.8, respectively. The films exhibit the ZnO (20-20) diffraction since ZnO films grow with strong (10-10) preferred orientation to its lowest surface energy [4].

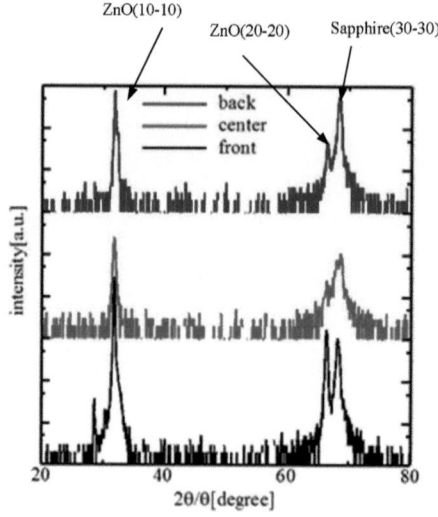

Figure.5.6. XRD spectrum for 8 L/min with 600°C

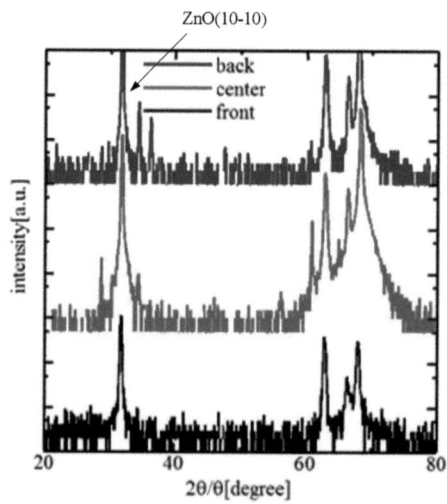

Figure.5.7. XRD spectrum for 8 L/min with 650°C

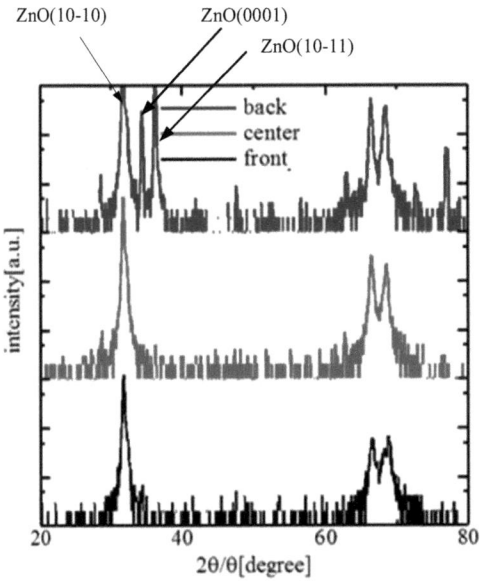

Figure.5.8. XRD spectrum for 8 L/min with 550°C

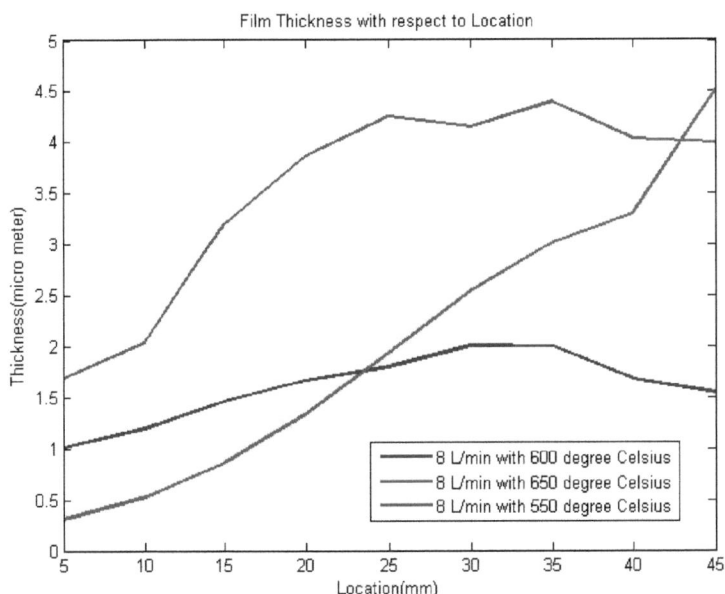

Figure.5.9. Film Thickness Measurement for (a) 8 L/min with 600°C (b) 8 L/min with 650°C (c) 8 L/min with 550°C

The film thickness measurement is done in Figure.5.9. According to these results, the best thickness condition is met at 8 L/min with 600°C. The summary table for first experiment is shown in Table.5.2.

Table.5.2. Summary for First Experiment

8 L/min	Film Thickness	PL (Deep Level Emission)	SEM	XRD
550°C	Small	Weak	Rough	Only m-Plane
600°C	**Large**	**Strong**	**Flat**	**Multi-peak**
650°C	Small	Strong	Flat	Only m-Plane

Figure.5.10 shows the PL Measurement for Front Side of the Sample on 8 L/min with 600°C, 8 L/min with 650°C and 8 L/min with 550°C. According to these results, the 8 L/min with 650°C case is the best condition among them.

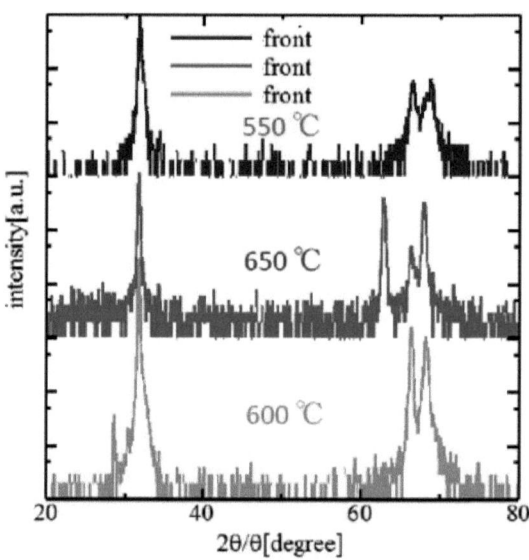

Figure.5.10. PL Measurement for Front Side of the Sample on 8 L/min with 600°C, 8 L/min with 650°C and 8 L/min with 550°C

5.2. Results and Discussions on Second Experiment

The deposition condition for this experiment is given in Table.5.3.

Table.5.3. Deposition Conditions for Experiment

Solution	Zinc Chloride aqueous solution (0.1 mol/l)
Deposition Temperature	600°C
Flow Rate	6 l/min, 8 l/min and 10 l/min
Substrate	m-plane Sapphire
Solution Amount	60ml
Solution	Zinc Chloride aqueous solution (0.1 mol/l)

Figure.5.11 shows the photographs of deposited thin films of ZnO at substrate temperature of 600°C with different flow rates for 6 L/min, 8 L/min and 10 L/min of nitrogen gas. The surface morphologies were characterized by SEM. In case of sample grown at 8 L/min and 10 L/min of flow rate, the front side of the sample has almost no grain boundary.

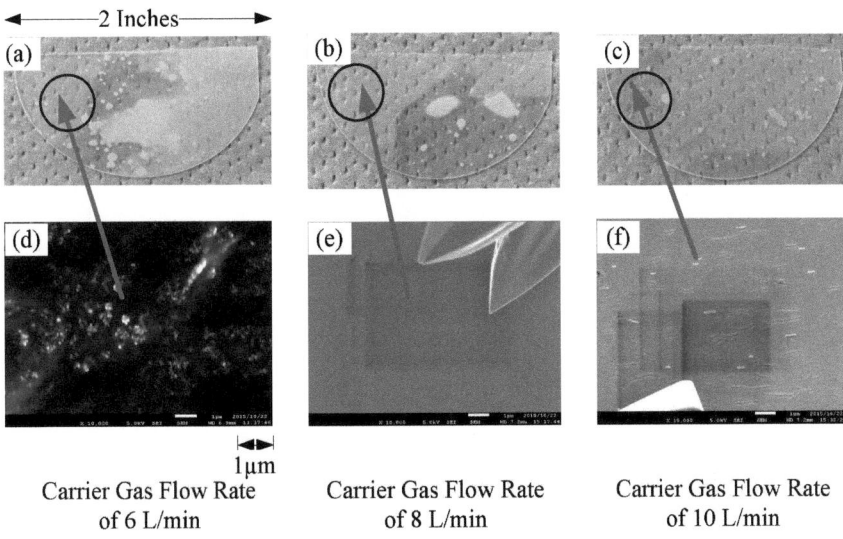

Carrier Gas Flow Rate   Carrier Gas Flow Rate   Carrier Gas Flow Rate
of 6 L/min   of 8 L/min   of 10 L/min

Figure.5.11. Photograph and SEM Images are Shown in (a)-(c) and (d)-(f),

Respectively

In order to characterize the optical properties of ZnO films deposited on m-plane sapphire, the PL measurements were performed at room temperature and the results are revealed in Figure.5.12. A weak near-band-edge (NBE) emission peak was observed at 3.26eV, while the deep level emission at 2.48eV which is related to oxygen vacancy [4-10].

In order to characterize the crystallinity of ZnO thin films, XRD analyses were performed and the results are shown in Figure.5.13. From the XRD spectrum for the sample grown with the flow rate of 6 L/min, c-plane ZnO (002) at 31.7°, r-plane (101), m-plane (002) at 66.3° and sapphire (100) diffraction peaks were shown. But the other two samples grown with flow rates of 8 L/min and 10 L/min possess m-plane (100), m-plane (200) and sapphire.

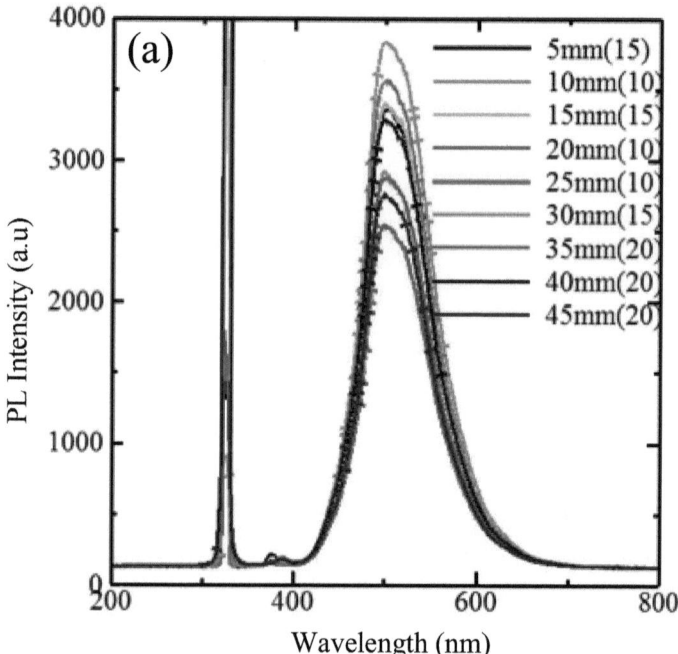

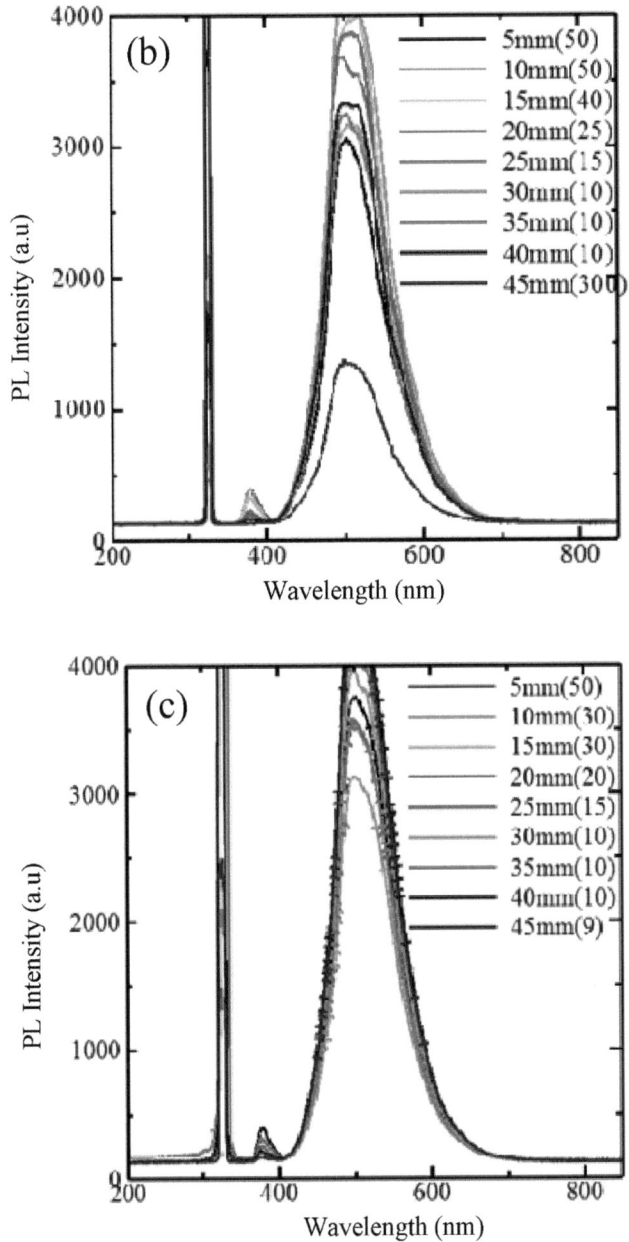

Figure.5.12. Photoluminescence Spectra for the Samples Grown with Carrier Gas
Flow Rate of (a) 6 L/min, (b) 8 L/min and (c) 10 L/min

Figure 5.13 (a) and (b) demonstrates the surface morphology for the position of 10000 μm and 20000μm of back side of the sample for 6 L/min with 600°C. There are high grain boundaries in this side.

(a) Back 10K

(b) Back 20K

Figure 5.13 (c) and (d) shows the surface morphology for the position of 10000 μm and 20000μm of center of the sample for 6 L/min with 600°C. This is the same condition of back side of the sample.

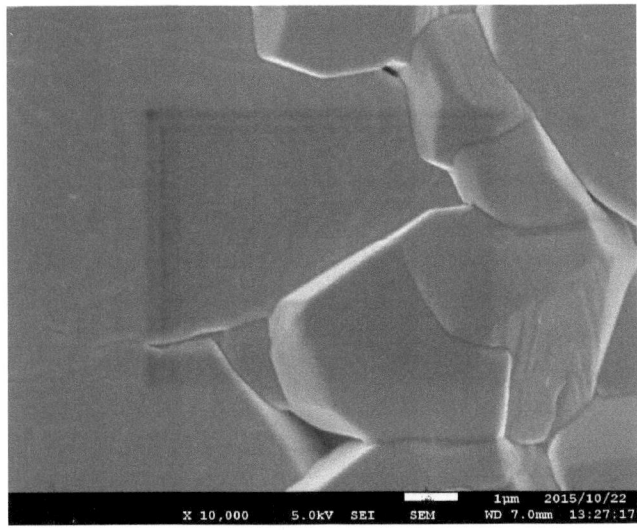

(c) Center 10K

(d) Center 20K

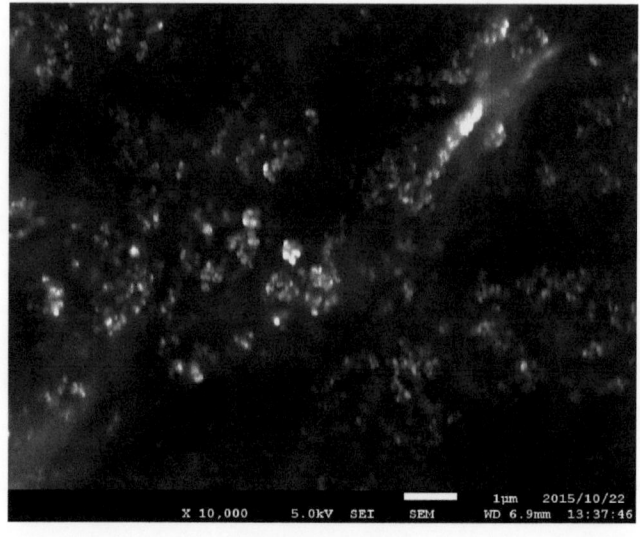

(e) Front 10K

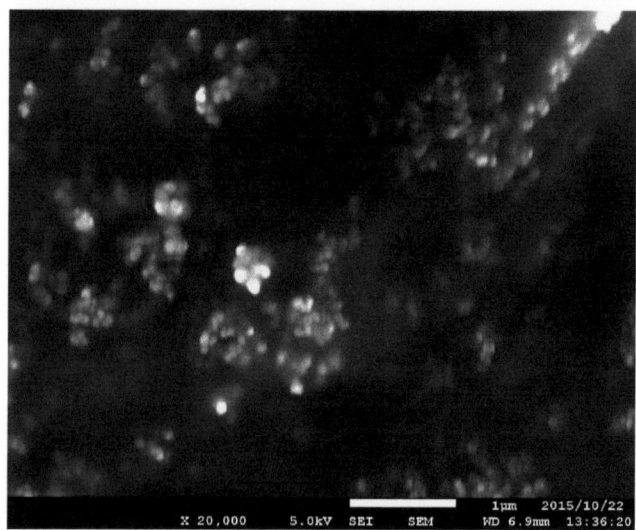

(f) Front 20K

Figure.5.13. Photoluminescence Spectra for the Samples Grown with Carrier Gas
Flow Rate of 6 L/min

Figure 5.13 (e) and (f) illustrates the surface morphology for the position of 10000 µm and 20000µm of front side of the sample for 6 L/min with 600°C. There are a lot of grain boundaries in this side.

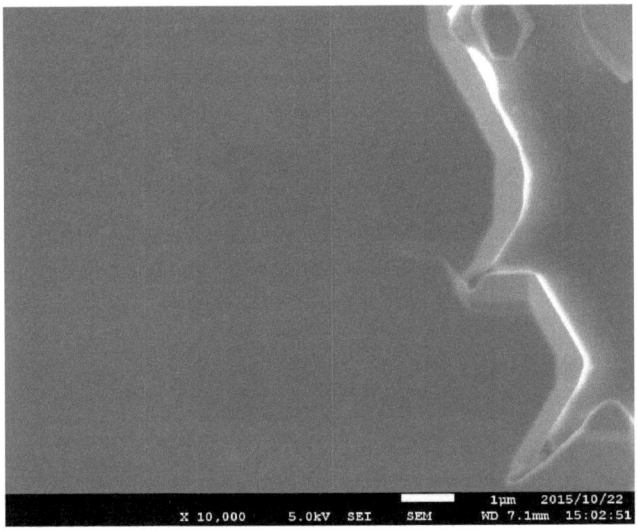

(a) Back 10K

(b) Back 20K

Figure 5.14 (a) and (b) confirms the surface morphology for the position of 10000 μm and 20000μm of back side of the sample for 8 L/min with 600°C. There are small grain boundaries in this side.

(c) Center 10K

(d) Center 20K

(e) Front 10K

(f) Front 20K

Figure.5.14. Photoluminescence Spectra for the Samples Grown with Carrier Gas
Flow Rate of 8 L/min

Figure 5.14 (c) and (d) proves the surface morphology for the position of 10000 µm and 20000µm of center of the sample for 8 L/min with 600°C. This is the same condition of back side of the sample.

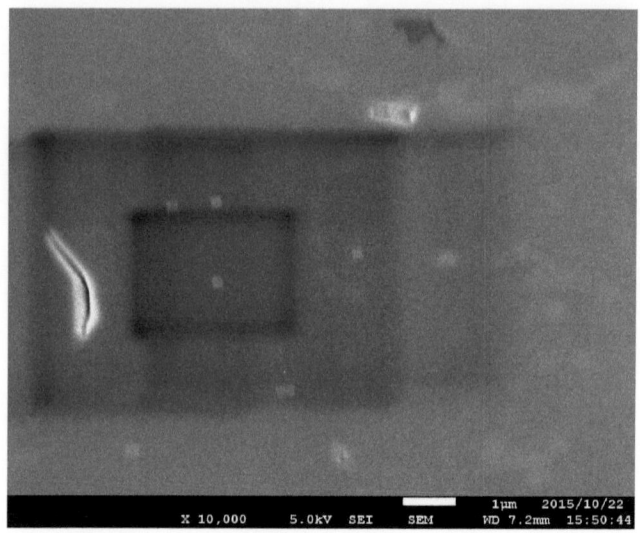

(a) Back 10K

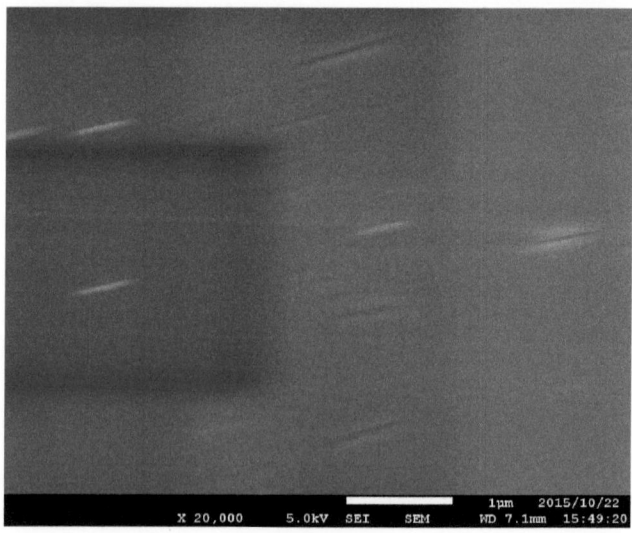

(b) Back 20K

Figure 5.14 (e) and (f) explains the surface morphology for the position of 10000 μm and 20000μm of front side of the sample for 8 L/min with 600°C. This is the same condition of back side of the sample.

(c) Center 10K

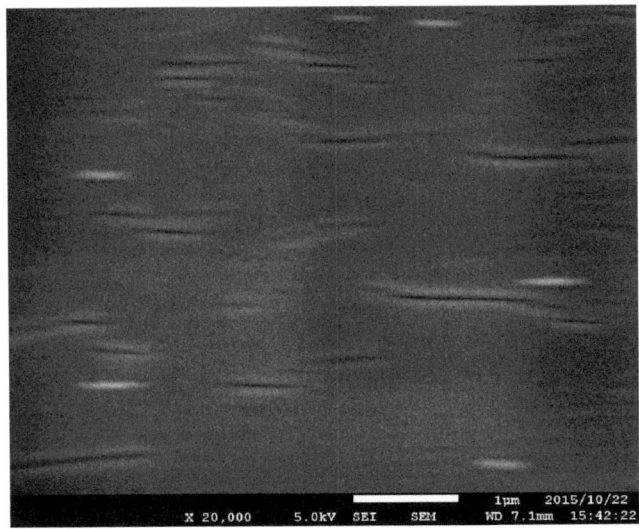

(d) Center 20K

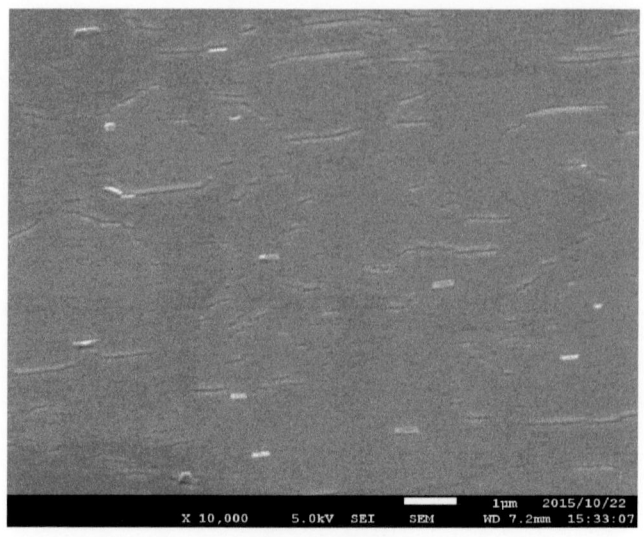

(e) Font 10K

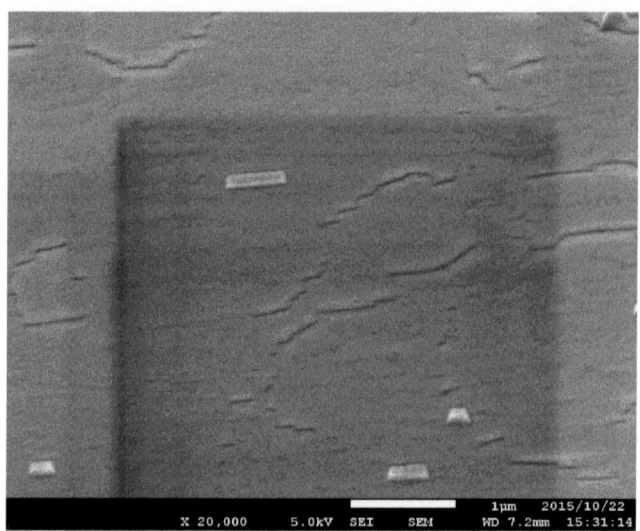

(f) Front 20K

Figure.5.15. Photoluminescence Spectra for the Samples Grown with Carrier Gas
Flow Rate of 10 L/min

Figure 5.15 (a) and (b) shows the surface morphology for the position of 10000 μm and 20000μm of back side of the sample for 10 L/min with 600°C. There are no grain boundaries in this side.

Figure 5.15 (c) and (d) demonstrates the surface morphology for the position of 10000 μm and 20000μm of center of the sample for 10 L/min with 600°C. There are no grain boundaries in this side.

Figure 5.15 (e) and (f) illustrates the surface morphology for the position of 10000 μm and 20000μm of front side of the sample for 10 L/min with 600°C. There are no grain boundaries in this side.

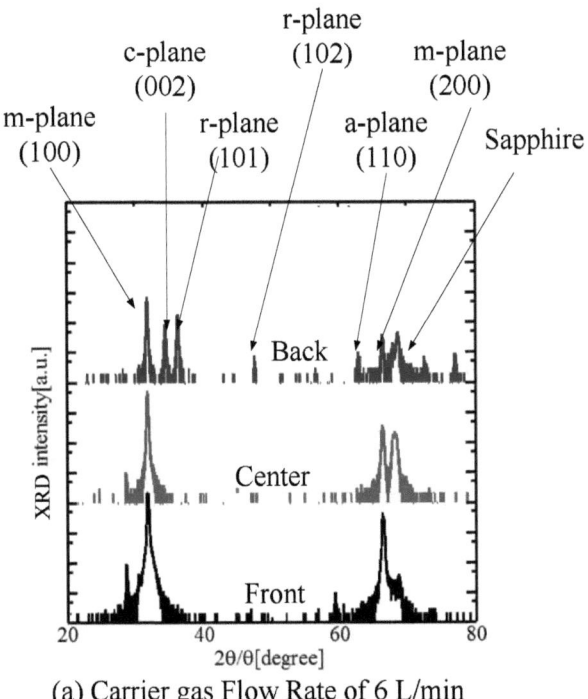

(a) Carrier gas Flow Rate of 6 L/min

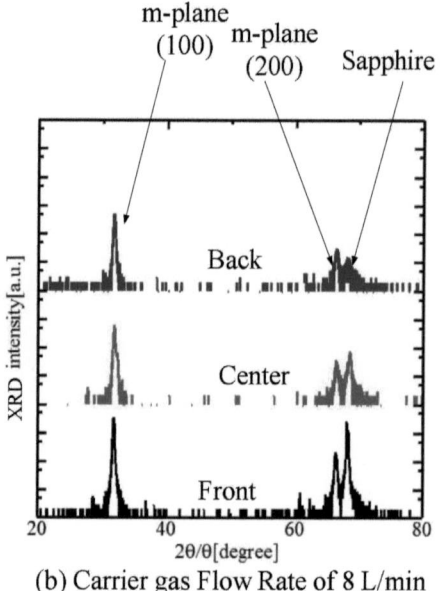

(b) Carrier gas Flow Rate of 8 L/min

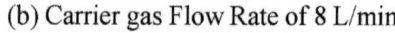

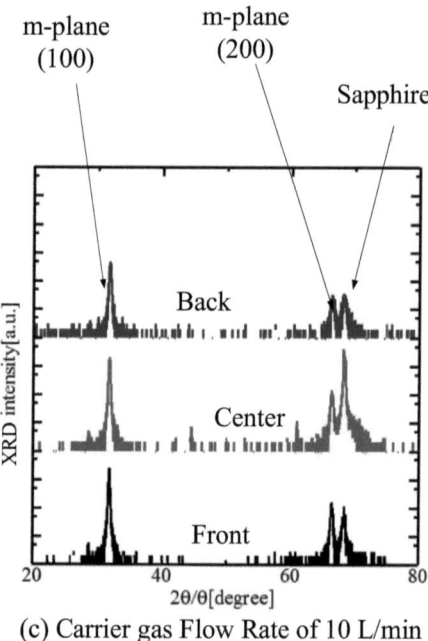

(c) Carrier gas Flow Rate of 10 L/min

Figure.5.16. XRD Spectra for the Samples Grown at Various Flow Rates

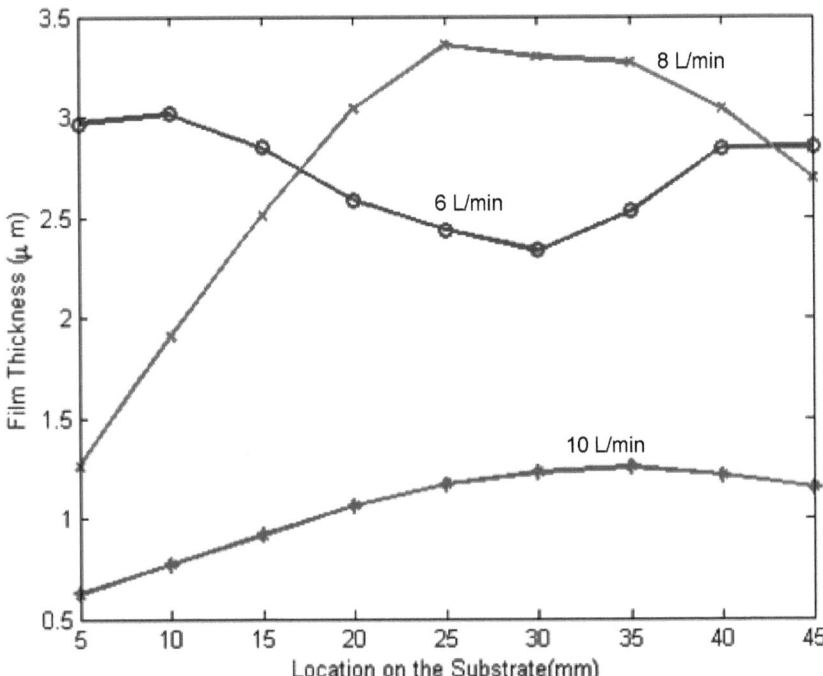

Figure.5.17. Film Thickness Measurement for the Samples Grown at Various Flow
Rates

The film thickness results are shown in Figure.5.17. When the clod mist come
into the furnace, then the temperature of the mist increase with the location on the
substrate. So, the thickness increases with the location because the thermal reaction is
enhanced by the mist temperature. When we increased the flow rate from 8 L/min to
10 L/min, the decomposition of the mist is decreased because the temperature of the
mist is not increased at the high flow rate. According to these responses, the crystal
growth condition for the sample grown with the carrier gas flow rate of 8 L/minis
was found to be optimal. The summary table for experiment is given in Table.5.4.

Figure.5.18 illustrates the PL measurement for front side of the sample at
various flow rates. According to these results, the 8 L/min with 650°C case is the best
condition among them.

Table.5.4. Summary for Second Experiment

600°C	SEM	Film Thickness	PL (Deep Level Emission)	XRD
6 L/min	Rough	Large	Weak	Multi-Peak
8 L/min	Flat	Large	Strong	Multi-Peak
10 L/min	Flat	Small	Weak	Only m-plane

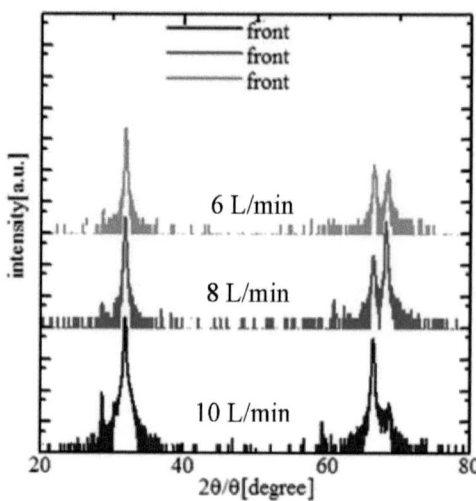

Figure.5.18. PL Measurement for Front Side of the Sample at Various Flow Rates

Chapter 6

Conclusion

In the research works, undoped ZnO grown on sapphire by mist-CVD technique has been characterized. The undoped ZnO on m-plane sapphire was growth with different substrate temperatures with 8 L/min flow rate of Nitrogen gas and different flow rate of carrier gas at 600°C. From the PL measurement, weak near-band-edge (NBE) emission peak was observed at 3.26eV while the deep level emissions were very strong. Film thickness decreased with the flow rate. According to this experiment, the optimum values for mist CVD techniques of undoped ZnO with m-plane sapphire are substrate temperature of 600 °C and the carrier gas flow rate of 8 L/min. In this work, we found the optimum condition for crystal growth of ZnO on m-plane sapphire and single orientation of m-plane ZnO was observed by XRD θ-2θ scanning mode.

References

[1]   Atsushi Tsukazaki, Akira Ohtomo, Takeyoshi Onuma, Makoto Ohtani, Takayuki Makino, Masatomo Sumiya, Keita Ohtani, Shigefusa F. Chichibu, Syunrou Fuke, Yusaburou Segawa, Hideo Ohno, Hideomi Koinuma And Masashi Kawasaki, Nature Materials, VOL 4, JANUARY 2005, pp-42-46.

[2]   K. Nakahara, S. Akasaka, H. Yuji, K. Tamura, T. Fujii, Y. Nishimoto, D. Takamizu, A. Sasaki, T. Tanabe, H. Takasu, H. Amaike, T. Onuma, S. F. Chichibu, A. Tsukazaki, A. Ohtomo, and M. Kawasaki, Applied Physics Letters 97, 013501 (2010); doi: 10.1063/1.3459139.

[3]   Kenji Yamamoto, Takako Tsuboi, Toshiya Ohashi, Takehiko Tawara, Hideki Gotoh, Atsushi Nakamura , Jiro Temmyo, Journal of Crystal Growth 312 (2010) 1703–1708.

[4]   J.G.Lu, T.Kawaharamura, H.Nishinaka, Y.Kamada, T.Ohshima, S.Fujita, J. Crystal Growth 299(2007)1-10.

[5]   Toshiyuki Kawaharamura, Hiroyuki Nishinaka, and Shizuo Fujita, Jpn. J. Appl. Phys. Vol.47, No.6, 2008, pp. 4669-4675.

[6]   N. Fujimura, T. Nishihara, S. Goto, J. Xua, T. Ito, J.Cryst. Growth 130(1993)269.

[7]   Kyu-Hyun Bang, Deuk-Kyu Hwang, Min-Chul Park, Young-Don Ko, Ilgu Yun, Jae-Min Myoung, Appl. Surf. Sci 210(2003) 177-182.

[8]   Shuji Nakamura, Masayuki Senoh, Shin-ichi Nagahama, naruhito Iwasa, Takao Yamada, Toshio Matsushita, Hiroyuki Kiyoku and Yasunobu Sugimoto, Jpn.J.Appl.Phys. Vol.(35)(1996) p.74-76.

[9]   F. Urbach, Phys. Rev. 92 (1953) 1324.

[10]   R.J. Elliott, Phys. Rev. 108 (1957) 1384.

[11]   C. Klingshirn, Phys. Stat. Sol. B 244, 3027 (2007)

[12]   G.F. Koster, J.O. Dimmock, R.G.Wheeler, H. Statz, Properties of the Thirty-two Point Groups (M.I.T. Press, Cambridge, MA, 1963)

[13]   Aoki, T., Hatanaka, Y., & Look, D. C., ZnO Diode Fabricated by Excimer-Laser Doping. Applied Physics Letters, 76 (22), 3257-3258, 2000.

[14] Y.R. Ryu, W.J. Kim, H.W. White, Fabrication of homostructural ZnO p-n junctions, Journal of Crystal Growth 219 (2000) 419-422.

[15] P. Zu Z.K. Tang, G.K.L. Wong, M. Kawasaki, A. Ohtomo, H. Koinuma and Y. Segawa, Ultraviolet Spontaneous And Stimulated Emissions Prom Zno Microcrystallite Thin Films At Room Temperature, PII: S0038-1098(97)00216-0.

[16] Y.R. Ryu, S. Zhu, D.C. Look, J.M. Wrobel, H.M. Jeong, H.W. White, Synthesis of p-type ZnO films, Journal of Crystal Growth 216 (2000) 330-334.

[17] Bixia Lin and Zhuxi Fu, Green luminescent center in undoped zinc oxide films deposited on silicon substrates, Applied Physics Letters Volume 79, Number 7 13 August 2001.